50 Questions e
Less Experienced Roller Flier

By Graham Dexter

Published by LALaS

1st edition published December 2013
ISBN 978-1-910148-00-6

Acknowledgements:

It is with great pleasure that I have produced this new book on Birmingham Rollers, and I believe it will be a great benefit to both the experienced roller fancier and the novice starter. However, I would not have been in a position to write such a volume with the help and support of many people. In the hobby, and over many years, I have had help, guidance, direction, criticism and been laughed at by many. Most of these fanciers, whether intentionally or not, have helped form my views, spurred on my enthusiasm, sharpened my insight and generally made me a better Birmingham Roller fancier. I thank you all! My special thanks go to those fanciers, some now passed, those that consistently and persistently did and do everything in their power to produce, train, and fly the very best Birmingham Rollers that they could. It may be unfair to name individuals particularly, but I feel deeply indebted to Dave Moseley and Bill O'Callaghan particularly, for without their help I may not now still be in the fancy. When my ebb was low, they are the two people who helped me most and I am truly grateful. This book is dedicated to them both.

I would like to acknowledge the many fanciers from the USA, Canada and South Africa for the interesting conversations about rollers that further informed my views. It was their interest in learning, and through various e-mails which sought information that inspired the production of this volume format. It is also necessary to publicly pay homage to the greats such as Bill Barratt, Bob Brown, Ernie Stratford, Ken White, Bill Pensom and of course Ollie Harris.....they were in one way or another my mentors and are the reason for all our successes.

I am indebted to my good friends in the hobby like George Mason (who I have quoted a few times in this volume), Les Bezance, Deano Forster, Morris Hole, Gordon Forbes, Johnny Conradie, and my late best buddy in the hobby George Kitson. Without these people my life in the hobby would have been much much the poorer. More recently I have had help and support from Jodie Rixon and the members of the Steel City Roller Club, the Yorkshire Performing Roller Society and the All England Roller Club. My personal thanks go to John Hall and Steve Taylor for continuing to keep the National Birmingham Roller Association going from strength to strength.

I would also like to thank those people who do things for the hobby in the background for very little recognition or thanks. For example, my proof reader and daughter Kez Gibbons, and all the administrators, treasurers, ring secretaries, trophy secretaries, fly managers and club members who make this fantastic hobby the great one it is!

Wishing you all the success that you earn and the pleasure you will derive from it,

Yours in sport,

Graham Dexter

General

1. Why do rollers roll?
2. What does a proper Birmingham Roller need to do to be worth watching?
3. Why keep rollers?
4. What's the best way to start?
5. What should I look for in a good roller?
6. What do I need to know to fly in competition?
7. What clubs should I join?
8. Where should I buy my stock?
9. What are the competition rules?
10. What do judges look for in the competitions?
11. How do I know when my rollers are good enough to compete?

Housing

12. What is the ideal setup?
13. How many pairs can I keep in a small loft?
14. How do I make my shed secure?
15. What are the most important things to remember when building a loft?
16. How much room does a roller need?
17. Do I need wire floors?
18. Why can't they fly loose in a large shed?
19. What are the best drinking vessels?
20. What's the best way to fly them in and out?

Flying and Training

21. How do I start to train my youngsters?
22. How can I settle old birds?
23. How long should they fly for?
24. My youngsters don't want to fly - what can I do?
25. My youngsters aren't rolling yet - what's the problem?
26. How can I get my youngsters to kit?
27. My birds fly too long - what can I do?
28. My birds don't fly long enough - what can I do?
29. Why do my birds keep rolling down?
30. My birds just fly around and don't roll - what can I do?

Feeding

31. What should I feed my youngsters on?
32. What should I feed my old birds on?
33. What's the best feed for breeders?
34. Does feeding really make such a difference?
35. What feed or seed will make my rollers roll more?
36. How much should I feed daily?
37. How important is grit?
38. Should I supplement the diet with vitamins?
39. What should I look for in the feed I buy?

Breeding

40. How many pairs should I start with?

1. Why do rollers roll?

A question every novice will ask, and every passing stranger looking up at a kit will wonder about. Despite lots of research into the possibility of epilepsy or some sort of lesion in the brain of rollers, even conducting experiments on them by treating them with anti-epileptic drugs, and numerous examinations post mortem, this question remains unanswered by science so far.

However the simple fact is that rollers roll because they have been genetically modified over the years to do so. Whether the induction of the first bird that tumbled over was a developed skill to divert attack, or to show off to a mate, or a genetic malady or mishap, man being man noticed this, admired it, and then harnessed the performance by skilful breeding. Not only did the early fanciers harness this potential, they enhanced and engineered it with patience and perseverance until it was perfected to their satisfaction. These days the modern fancier is the beneficiary of such dedication and endeavour of our predecessors.

My view is that a behavioural trait such as this is similar to how some human families carry abilities over generations in fields such as mathematics, art, music or particular manual skills such as athletics. The difference is only in the fact that human families have no external force choosing mates and putting together breeding programmes, so sometimes well-developed skills and abilities in human families are lost or diluted.

So let us keep the faith and ensure that performance so perfected is maintained. The standard so well described by Pensom:-

The Birmingham Roller should turn over backwards with inconceivable rapidity for a considerable distance like a spinning ball.

2. What does a proper Birmingham Roller need to do to be worth watching?

Although this seems a simple question it does defy clear description, despite Pensom's definition. There are a few criteria that help the discerning fancier ensure he/she is keeping the best pigeons they possibly can:

- The roller should roll over backwards holding its wings up to a parallel point at the top of its axis, and on the wings on the down stroke should also finish when parallel. This gives the illusion (if fast enough) of a smooth spinning ball. When seen from below it may present as a spinning ‘H` or if the wings go slightly past the parallel position it may look like an ‘A`. When the wings do not reach a full parallel point before commencing the downward stroke it may appear like an ‘X` from below. For most novices the most important thing to note is that when the bird is rolling and viewed from the side at a right angle, it should give the impression of either a solid ball or a ball with a hole through the middle.

Pensom’s ‘considerable distance’ should be long enough for the viewer to see it start, notice its shape, speed, and style, and whether the bird stops correctly. If the roll is too short the viewer will not have enough time to do all this. I do not like the current estimate of length employed by most fanciers, often you hear them say the roller is ‘only ten foot` or ‘at least 30 foot`, these estimates are very subjective and there is no common understanding of these assessments so they tend to become meaningless. It is really difficult to estimate distance when there is nothing in the sky to measure against. I know that whenever I have had a roller roll during landing, a very short spin can actually take up at least 20 foot. I know this as I could use my landing pole, which was 25 foot high, as a measure, and even the short spinners sometimes started above the pole and still hit the top of the shed! I prefer to measure the depth of the roll with a stop watch or at least an estimate of time in seconds. A short roller will roll for a second to a second and a half, medium depth rollers roll between 2 and 3 seconds, and deep rollers roll for 3 seconds or more.

- The proper Birmingham Roller should
  - start cleanly and sharply in the roll,
  - roll in a straight line,
  - show no change in speed or style throughout the roll
  - stop cleanly, facing the same way as the kit.

- The bird should be capable of rolling frequently (the old standard used to be once a minute, but perhaps then a better standard is at least as often as the rest of the team), and should be able to roll throughout the duration of the fly.

3. Why keep rollers?

Rollers are one of the cheapest and most rewarding hobbies I have ever heard of. They take little time if that's what you have, and can be completely absorbing if you have plenty of time to give them. A dry shed, clean water, ample grit, nutritious food, and your rollers will repay you with endless hours of joy and pleasure. Only when you become fanatical and obsessional about them will they break your heart and frustrate you to death!

4. What's the best way to start?

Find a local fancier, join your local club, watch as many competitions as possible, travel and see as many rollers and rollermen as possible, and train your eye to ensure that you know what to look for and what is most attractive to yourself. Once you have done this, concentrate on finding the person who has the sort of Birmingham Rollers that please you most and take advice, stock and learning from that person. Listen to everyone, and decide for yourself.

5. What should I look for in a good roller?

All of the things stated in the answer to Question 2. And that the bird is healthy, has been well bred, is from an established family, and is from a fancier well respected in the hobby. The bird itself should look alert (even perhaps a little wild in the eyes), with good feather condition - the bird should slip through your fingers smooth as silk if it is in good feather condition. It should be what is known as 'apple bodied` i.e. if you look at it from the side you could see the shape of an apple tilted at about 45 degrees, the stalk pointing to the top of the bird's head and the wing butts of the bird forming the top part of the apple and the rest of the apple forming the keel and finishing at the base of the tail. The bird should weigh about 7-10 ounces and look compact, not long keeled and gangly. This sort of shape will look more graceful and streamlined in the roll than any other shape and size. Although the colour is not important, when selecting stock it can be useful to have a range of colours. This will help you identify them better in the air, and leave you less bored in 10 years' time!

6. What do I need to know to fly in competition?

Very little really. You will need a team of birds that have a recognised ring on them and the requisite number - usually 15 to 20 for the club rules of any local or national club. More importantly, you will need to have the courage to listen to the judge's verdict, be prepared to receive constructive criticism, advice, and perhaps even a little ribbing from some of the more vocal members of the entourage. Train your birds so that when they leave the loft they will take to the air without too much prompting and that they will land on the loft top (not on the house roof) and go in without too much of a problem. The more you can demonstrate that your kit is under your control, the less ribbing you will receive from any spectators.

7. What clubs should I join?

Always join your local club and the national organisation when one is available. In England the All England Roller Club is essential to join, and it is usual that any national club such as this will be able to put you in touch with your local club and local fanciers. Always join a club that deals specifically with Birmingham Rollers of performance, and avoid clubs which specialise in the show side of the hobby. I have nothing against show clubs, but they must not be confused with clubs that deal specifically with the proper performing Birmingham Roller. Over the years I have heard complaints from roller fanciers that their rollers don't perform as I have written about in my articles and books. This has nearly always been because they have bought birds produced for the show pen and not for performance in the air, or they have bought birds from trade magazines from unscrupulous fanciers more interested in money than in the hobby.

8. Where should I buy my stock?

Unfortunately there will always be unscrupulous fanciers who will want to take advantage of the novice, but joining a bona fide club should insulate you from most of these characters. Even in clubs, sadly, there will be those that want to rip novices off, and sell their sub-standard stock at premium prices. Personally I find this appalling, and some of the current prices for birds of dubious note are far from what I would call value for money. So shop around and don't be misled by the first offer of help from club members. Have a good look around, by all means buy a cheap kit of youngsters to practise on, learn how to fly and control a kit of rollers, but when acquiring stock to breed from, great care must be taken. Always buy stock from someone who has been competing and doing reasonably well. This is not necessarily the National winners, but someone who is consistent at local level. This person is more likely to have a reasonable price in mind, and someone who might be interested in helping to mentor you.

If you have been around on the flying competitions and made yourself known and popular, then often fanciers are willing to help out and let you 'test out' their birds' breeding potential for them. You can be doing them a favour, they may have a few flying birds that they are tempted to use in the stock pen but don't have the capacity to put extra pairs together - this is the ideal for you. You have the benefit of some unproven potential and the interest of the fancier supplying the stock. He or she may want to do a deal if anything spectacular comes out of the experiment, but you will still have a greater chance of some great birds without having to pay the earth for them, so it's definitely worth thinking about.

Those of you for whom money is no object, then I would suggest you go to the fancier who has been in the top ten of the national competitions for the last 10 years. From this person you will be able to buy a strong family of birds which are likely to breed true, and because they value their reputation you are unlikely to be sold substandard stock. Always ask the question 'so would you breed from this bird yourself?` If the answer isn't a definite yes, keep your money in your pocket and go elsewhere.

9. What are the competition rules?

Competition rules vary according to the custom and tradition of the various locations, however, there are some commonalities which I believe are underlying principles of most Birmingham Roller competition rules:

- The number of birds in the team is set up by each club but generally the tradition is for teams to consist of 20 birds. In order to help engage less established fanciers the English National Competition allows for a number between 15 and 20 to be flown. Most clubs in England now follow this protocol.

- There will be a time limit on the judging - this is usually 20 minutes on the clock. However, most clubs now allow for no penalties to apply once 15 minutes on the clock has expired, so a whole team could land after 15 minutes without penalty.

- Most competitions give the competitor a short amount of time (usually 10 minutes) to allow the kit to form before electing for them to commence judging. After this time the judge will start judging in any case.

- The judge will allow no interference with the birds once on the clock, and he or she will be in control of the clock once on. For example, there may be local rules which allow for the clock to be temporarily suspended for eventualities like the birds flying too high, or going out of sight, or a predator attack.

- There may be penalty points for birds landing before the stipulated time, rolling down or alighting anywhere other than the loft top. Some rules stipulate a disqualification if a number of birds land before full time, or a specific number of minutes.

- Points are awarded for the number of birds spontaneously rolling together, and additional points for the quality of roll on display. For example, in the English National rules 200 points can be awarded for the quality of roll in the team. Some judges consider this as 10 points for each bird that rolls with excellent quality, others simply follow a 'gut reaction` to award up to 200 points for the overall quality of the team. Points are awarded on the basis of single points from 5-9 birds rolling together, double points from 10 -14, triple points from 15-19 and 200 points for a full turn of 20 birds. Alternatively, in the World Cup rules, points are awarded for spontaneous rolling and then this score is multiplied by both quality and depth assessments of between 1 and 2 - thus a raw score of

100 for breaks can be increased by say 1.5 for depth and say another 1.6 for quality leaving a final total of 240 (100 x 1.5 = 150 x 1.6 = 240)

- In some competitions points are also awarded for how closely the team kits together, in the UK this is generally 1 point for each bird in the kit, thus maximum kitting points would usually be 20 points. And alternatively in some competitions - notably the World Cup - there are no points for kitting, but there is a rule that when 2 or more birds leave the kit (other than when returning to the kit after rolling), the judge suspends judging until the birds return or at least there is only one out of the team.

10. What do judges look for in the competitions?

It will of course depend on the judge, but essentially the judge is there to find the best team of Birmingham Rollers in the competition. He or she has to do this within the confines of the rules and in the process is also going to have to evaluate your team and give you some feedback. Any good judge will want to see a team of rollers demonstrate that they have the required standard of quality performance, that they stay together as a team, and that they work together as a group of no less than 5 birds together. Some judges are very strict when it comes to quality, others seem to be very concerned with how well the team kit together. Others are very particular as to the absolute spontaneity of the performance, and yet others will not award points unless they are convinced your rollers are rolling with sufficient depth to classify them as Birmingham Rollers. Ideally your team should be able to fly together, roll together and roll with sufficient quality to score from any judge. If your team achieves all the above criteria and roll for at least 1.5 seconds, any judge appointed should score them well.

11. How do I know when my rollers are good enough to compete?

As previously stated your team should kit well, i.e. they should stay in a compact group until they make a break (more than 5 rolling together). At this point it is possible the team will split up a little, but they should all be back together within a few seconds. The amount of time it takes for the birds to regroup will often depend on the depth of the deepest rollers in the team, this is why some of the most successful fanciers try to have birds in the team that roughly roll at about the same depth and have a few shorter rollers to hold the nucleus of the team together. Your birds (or at least a good percentage of the team) should have enough quality to demonstrate that they are Birmingham Rollers according to Pensom's definition. They must spin over backward so fast that the individual somersaults cannot be counted by the naked eye, and do so for at least 1.25 seconds (and a lot of judges need 1.5 seconds) with a graceful style. When I say graceful I mean that when watched from the side, front or below they appear to be a round ball, with no jagged edges sticking out. There should be no jerkiness in the roll, and a straightness in their trajectory. The speed of somersaults should not change during the delivery of the roll, and the finish should look like they have cleanly snapped out of the spin rather than slipped out to one side. I am of course at this point describing the perfect Birmingham Roller and there are few teams (even with the great and legendary fanciers) that achieve such quality in all of the individuals in the team, but the higher the percentage of 'proper` rollers in the team the greater the overall 'quality` of performance the team will convey. The shortest answer I could have given to this question is ....you really won't know until you enter and have feedback from an established judge!

12. What is the ideal setup?

Everyone will have their own preferences, yet there is something to be said for the traditional design. It is important to have flying kits housed in compact (not cramped) conditions as this helps the team to do three things:

- Fly together in a compact style.
- Avoid using excess energy free flying.
- Avoid pairing up and producing eggs. A hen in the flying team that pairs up and begins to produce an egg is at grave risk of injuring herself when rolling. The egg sac is stable when empty but not stable when an egg is being formed and especially when frequent and fast rolling prevails.

For twenty to twenty-five birds, a square yard (or metre) is required with at least 2 extra perches than the number housed, and a wire floor for droppings to fall through.

The wire floor (2" x 2" mesh) is necessary to prevent fouling of food left uneaten, and to ensure the least amount of cross contamination from droppings can occur. Worms, canker, salmonella and coccidiosis, which are common diseases in rollers, are all transferred from one bird to another through contaminated food and water. Thus feeding trays and water vessels should be cleaned and if possible sterilised frequently.

It is essential that water is kept clean and refreshed regularly to prevent this cross infection, and some fanciers use hypochlorite solutions (bleach) in the water (at low concentrations) to avoid such cross infection. There are preparations on sale from pigeon suppliers to effect this practice. If loose droppings occur due to the use of antibiotics or in some forms of 'young bird sickness', preparations of dried live yoghurt can be added to the water to help restore the natural balance of bacteria in the birds' digestive systems.

Breeding lofts can be of any size, but ideally each pair should be provided with 2 nest boxes to avoid too much fighting unless the pairs are in separate boxes and locked in during the breeding process. The latter is only recommended to ensure that youngsters are 'legitimate' and not fathered by other cocks; however this will inevitably slow the production of youngsters. This is because the general 'breeding activity' (billing, cooing, chasing, tail spreading and treading) is the natural stimulus for breeding to be prolific; locking pairs in boxes tends to de-energise this process. See the illustration of kit box and general loft layout, provided by Les Bezance many years ago.

Breeding boxes should be a minimum of 18" long, 15" deep and 12" high. This space needs to be at least doubled if the pairs are to be locked in for the duration of the season. Clean water and food needs to be supplied continuously to these locked compartments, and regular cleaning out is required. It is recommended that if pairs are to be confined they should be given liberty in the loft or aviary as often as possible, but obviously without any other cocks having access at the same time.

Perches can be of many different types, but Bill O'Callaghan's design of box perch is the desired one for me. This is the typical box design of interlocking plywood strips constructed so that the sections are 10" wide, 8" high and made of strips 2¾" strips. Although this ledge seems rather narrow to us, the birds appear quite happy with them and as they have to turn around to defecate they tend not to foul other birds in the lower perches.

Diagram of loft suitable for Breeding and Flying

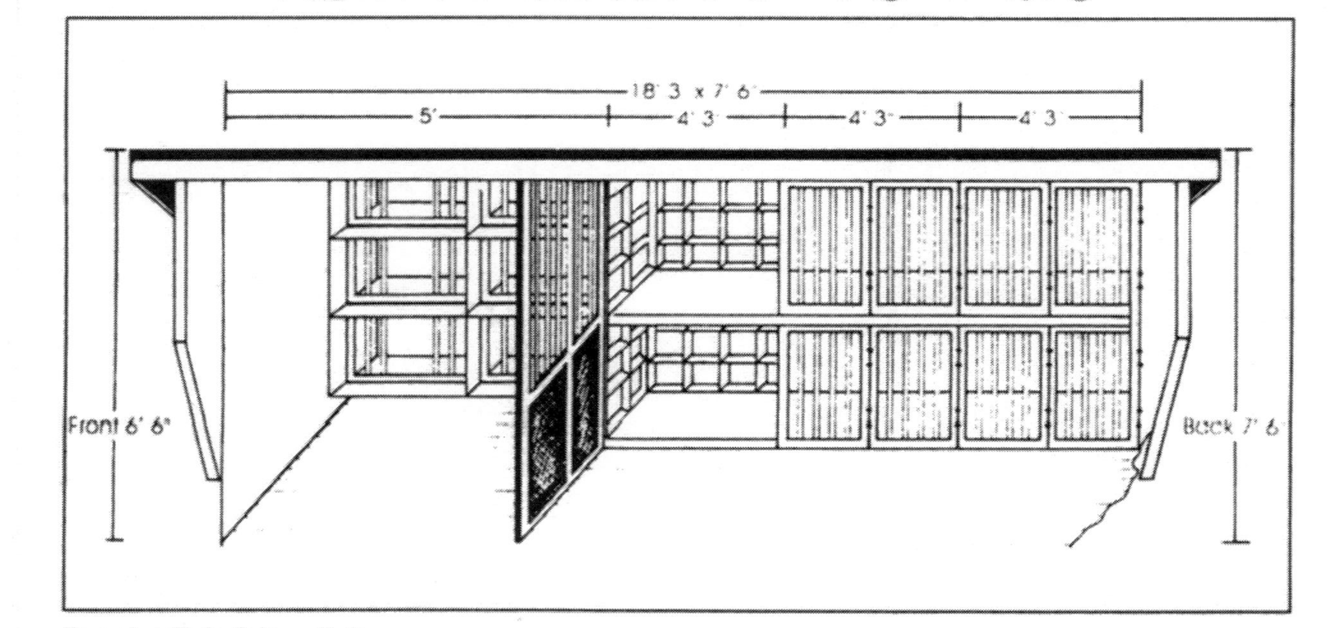

Breeder 7 ft 6 in x 5 ft.
Ideal for 6 pairs.

6 boxes with front shelf with no other perches except the front of the cages.

6 Kit Boxes 4 ft x 3 ft to hold 21 birds in each.

Either box or 'V' perches.

Double-hinged doors on Kit Boxes.

13. How many pairs can I keep in a small loft?

A breeding loft with a floor area of 6 foot x 4 foot excluding the space taken by the breeding boxes can happily accommodate 6 pairs of rollers. More pairs than this will increase the level of illegitimacy (cocks treading hens other than their own). In cramped conditions the rate of illegitimacy can be as high as 11%. As for a flying loft, I have in the past constructed kit box lofts which are simply one square metre on legs constructed with wire floors and plenty of ventilation at the top back and front bottom - this allows for a flow of air through the box without allowing wind and rain to trouble the occupants. See the illustration below.

Flying Pen

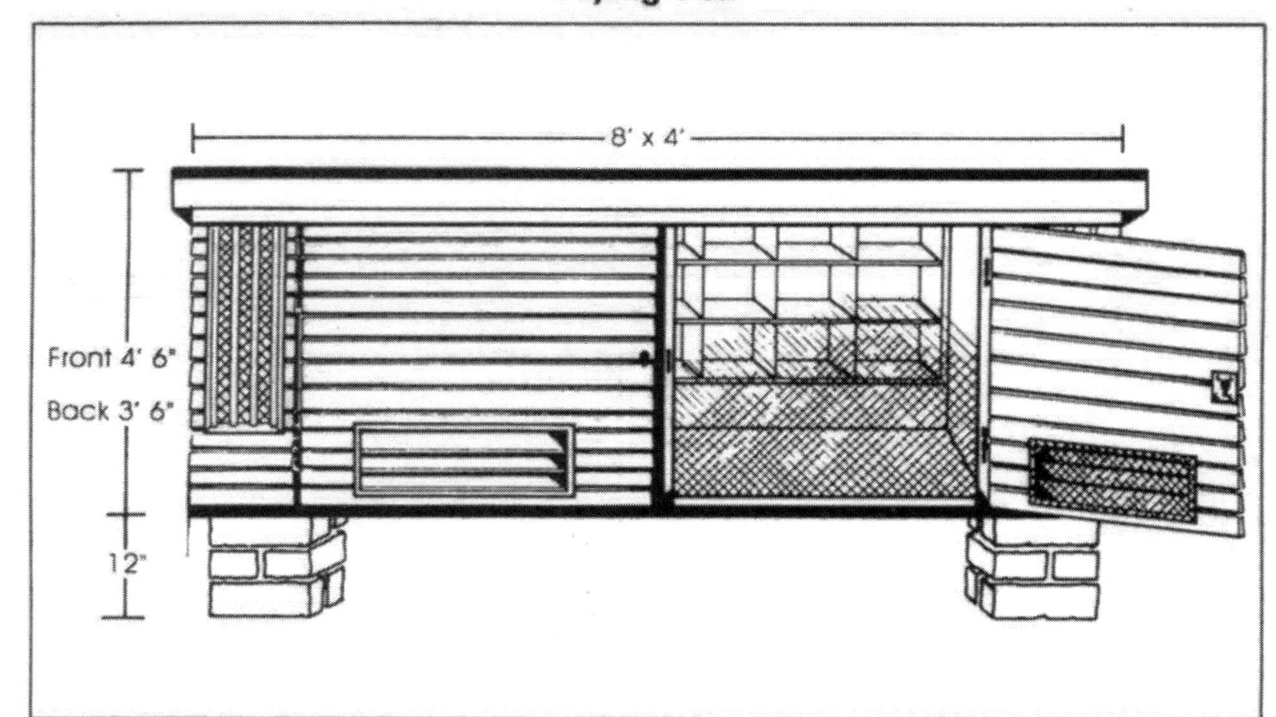

To hold 2 Kits, 6 in. louvre vent through the full length of the back. All dowels and louvres covered with fine mesh. Also inserts can be made for door openings in fine mesh so doors can be kept open in warm weather.

Plan view of the Author's Flying Loft

Plan view of the Author's Breeding Loft

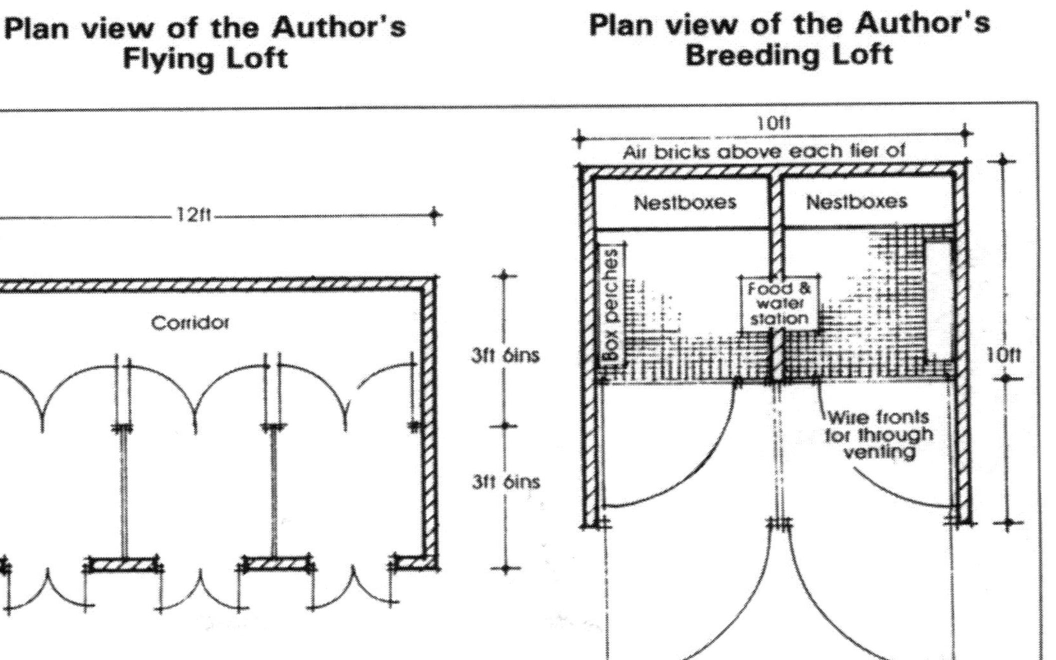

Note the corridor behind the Kit Boxes for easy access. This allows the owner to chase out the birds very quickly.

Note the wire floors, double doors to prevent escape of 'prisoners' and 2-section partition to allow for separation of sexes out of season.

14. How do I make my shed secure?

Making a loft secure is much harder than one imagines. In the UK we have many human predators (thieves) than regularly steal top quality stock. Making a loft impregnable is impossible, but making it difficult for burglars is important. Pressure mats, security lights, double skinned walls and steel bars can be helpful. Door and window alarms are often a deterrent. Expert advice from security specialists will be helpful. With a bit of luck this might keep most humans out, but complacency will often catch you out - a few years ago one unfortunate person boasted of his impregnable loft only to discover his stock taken one morning. The thieves had seen his boast as a challenge and had entered his loft through the floor by using trolley jacks and brace and bit drills!

Cats, rats, mice, weasels, hawks, foxes, mongooses, snakes and other such predators also need to be kept at bay. This requires that the loft is serviced by good ventilation with small wire netting backing. Doors and windows need to be tight fitting as it is amazing how small a space some of these creatures can steal through. One year in Portugal I had eggs disappearing from the nests, and after dismissing the possibility of broken eggs, I was at a loss. Eventually I left a poisoned egg on the floor of the loft to discover it missing the next day....no further eggs went missing after that. I concluded that a small snake was getting in and taking the eggs. I never found out how, but had no further trouble. It is necessary to check the loft corners and any covered areas (nest boxes and storage areas) from time to time, as mice and rats can gnaw their way through even thick wooden sections if undetected for long enough. Lofts should be at least 1 foot clear of the land they are erected on so that these creatures can be clearly seen if they are ever under the floor. Wire netting should also be placed around the bottom of the loft so that other creatures such as cats cannot hide under the loft waiting for the opportunity to grasp an unsuspecting roller pecking around the perimeter.

15. What are the most important things to remember when building a loft?

Remember that rollers are pretty hardy creatures, so they thrive on fresh air and don't dislike cold. However they do not like draughts, damp, and parasites. So ensure that fresh air is available in your design - lots of wire netting behind secure windows. Have a flow of air from low front to high back but no ill-fitting panels that let draughts through. Ensure that rain and damp cannot penetrate the main housing areas. Construct the loft in materials that can be disinfected and easily cleaned. Plywood covering tongue and grooved construction is useful as it can be easily scraped and sprayed with disinfectant and anti-parasitic solutions and tends to dry out fairly quickly. Lofts constructed of brick or breeze block are less easy to clean but can be sprayed to disinfect and de-parasite. Cement floors have the advantage of deterring the entry of predators and can be cleaned and disinfected easily.

Bearing in mind that you the fancier will have your own abilities and disabilities, design your loft for personal ease. For example ensure that the top corners of the kit boxes can be reached without having to stretch too much. You will want to be able to catch that bird, and you can bet your life it will be just out of reach if it's possible to be! Design your nest boxes so that they are easy to clean, and that cleaning floors can be done without too much bending or kneeling. This might mean a little wasted space where you could have fitted that extra nest box or kit box, but in the long run - especially when you are feeling less energetic, have some malady, sprain, strain or just plain back ache, the loft is still manageable.

16. How much room does a roller need?

As previously stated 20 -25 rollers require 1 cubic yard or metre of space if exercised regularly. If they are to be confined they will need more. I would recommend 20 birds will need 3 cubic metres when not given liberty with access to an aviary when possible. It would be important that cocks and hens are separated in this event, otherwise you will have them attempting to pair and produce without proper provision. Having said this you should expect to have the hens pair to each other and still produce eggs (obviously infertile). Although I like to give my rollers access to grit as this helps them get the best value out of the grain you feed them, it is wise to omit this when hens are segregated as grit will encourage them to pair and lay.

17. Do I need wire floors?

As I suggested in the 'ideal setup', kit boxes are better with wire floors as this limits the amount of contamination from pecking at droppings, it also acts as a slight deterrent from flying birds pairing and the consequent laying of eggs. Having said that solid floors have been quite successful for many fanciers over the years, ideally these will be cleaned regularly and a floor dressing should be used. Sand, sawdust, or a combination of the two is suitable, but also mixing this with some disinfectant dressing will limit the possibility of cross infection through droppings. There are many proprietary brands of floor dressings on the market, and cheap varieties of 'cat litter' can be used for economy. Whether wire floors or floor dressing is used it is still essential that a large area of clean flooring or feeding tray is used to feed the kit. I suggest a large area, as when small trays are used you may find that some of the smaller birds and slow eaters will suffer from being excluded by the stronger more aggressive birds in the team.

18. Why can't they fly loose in a large shed?

Well of course they can, however they will not be quite so keen to fly when released, and they will pair up, and you will find eggs being laid in corners (even if only hens are housed together). As mentioned before a roller rolling its egg sac out is not pleasant to watch, painful and often fatal for the bird, and will certainly leave the lucky survivor infertile.

19. What are the best drinking vessels?

There are numerous vessels available commercially; I favour anything that is easily cleaned and disinfected. Most are designed so that the birds cannot easily foul the water but if a sand or sawdust floor dressing is used this can easily contaminate the water. Hence daily replacement of clean water in the vessels is necessary. I have an automatic water system in my loft which is fed to each cup via a plastic tube which is supplied with water from a header tank. Ideally these are situated under a 'V' perch so they cannot be fouled, and have the advantage that they remain empty until a bird presses the stainless steel pin in the plastic cup to release water. With this system fresh water is always available, although the plastic cup does require a wipe with an absorbent cloth from time to time.

20. What's the best way to fly them in and out?

A large open door is definitely the best way to fly rollers out and to get them quickly back in. There are all sorts of pigeon traps on sale commercially, but these are unnecessary. When birds are trained correctly the main door to the loft is perfectly satisfactory. Occasionally one may have difficulty with a youngster that is dazed from its rolling and may linger on the loft top, or an old bird that has had a bump may be reluctant to come in straight after alighting. Sometimes if a team has been overfed the day before they may linger and want to bill and coo, spread and tread, and generally misbehave rather than go straight in, but these should be the exceptions rather than the rule. For these eventualities one can put a simple `bob' hole in a strategic place in the door or wall of the loft, giving access either to the main loft or directly into the kitbox. My recommendation for this would be a simple round hole of 6" diameter covered with a flap that drops down to form a landing perch, itself as small and fragile as possible to avoid it being used by heavier predators such as cats. This then can facilitate the late or lingering bird to enter the loft without allowing any of the more obedient birds to come out after having eaten. In *Winners with Spinners* there is an illustration of a device `a coffee tin trap' which does this job remarkably well (see the illustration below). However if these devices are to be used all the birds will at some time have to be trained to use them.

21.How do I start to train my youngsters?

From the first day that they are removed from their parents, feed them twice a day. Remove any uneaten food from them, and before and during their feeding use an aural stimulus to condition them to the feeding. This aural stimulus can be anything convenient to you, but traditionally this is the shaking of the food container (usually an old bean tin). Other fanciers have used whistles, silent whistles tuned to the correct frequency, or voice calls. This is extremely important in order to have some control over the birds when they first begin to fly unrestricted.

Start training youngsters as early as possible and by 4 weeks old they can be shown the world outside of their loft. A wire cage where they can see the front of the loft needs to be placed in a safe place perhaps 6 to 10 foot away from their home. It is important that this cage is not on top of the loft, as this only provides the birds with a view of the loft top and the surrounding environment, not the sight of the front of the loft which you will be expecting them to return to. Preferably this cage should have a view of the kitbox they are housed in through the open door of the loft, failing this, the floor area where they are usually fed. If you are bothered by neighbouring domestic cats it will be necessary to construct a stand for the cage to be secure again such tormentors. A steel pole constructed like a hangman's scaffold would be a good design.

About 6 foot high would be high enough to deter most cats. Wooden construction is not appropriate as cats are very good at negotiating wooden structures; steel ensures that their claws are redundant. Failing this it would be important for someone to constantly watch the cage to ensure that the birds inside were not frightened. For the first few days, the birds could be fed and watered in the cage and placed in it for a few hours at a time. This should settle them to the cage, and later it may be helpful that they should remember that this is a safe place. They should always be taken back to their kit box after such excursions. After several days and frequent excursions to the cage and ensuring that they are hungry (perhaps remove food from them for at least one day), entice them out of the cage with your aural signal to either their kitbox or the feeding area, whichever is their usual feeding place. In the event that one or two fly up rather than down to the feeding area, the cage as a place of safety will provide some backup for the bird to settle on rather than fly up and be lost. Continue this routine for a few more days, getting them to fly from the cage to the loft.

Once this has become habitual, then the cage can be dispensed with and the birds allowed to leave the kit box and fly out, perhaps even up to the cage and back. Once the youngsters get used to this they will begin to land on the loft top and come in and out on the sound of the corn tin being shaken (or whatever aural signal you have trained them to). After about a week of this the birds will inevitably want to fly and a few will take off. Signal them down as quickly as possible so that they land to the loft top. To allow them to fly too much or too high before signalling them back may lead to their misbehaviour - landing elsewhere, or flying too high and becoming lost. Take time to get them acclimatised to this striking up and landing. Once you are convinced that the youngsters are happy to consistently land on the loft top, then more liberties can be taken, and the birds not called down too quickly. If the birds are healthy and well bred, the youngsters will soon begin to kit and fly for a few minutes, and gradually the flight time will get longer as the birds build up strength and fitness. If for any reason the birds do not take to the air spontaneously then it may be necessary to encourage them. Many fanciers use a flag and encourage them this way, I prefer to let the youngsters feel safe on the loft top, and if encouragement is necessary I prefer to basket them and release them from a few yards away, perhaps up to 100 yards away, and this way they are encouraged to fly but without losing their security on the loft top and feeling safe to land without fear.

22.How can I settle old birds?

Old birds are a bit more difficult to settle, and I have to say it is likely that you may lose some in the process of trying to settle them. However these are the main points that should at least reduce your losses:

- Ensure that the birds spend some time in a cage where they can safely see the front and top of the loft, and have the best view of the neighbourhood.

- Feed them grains that will slim them down and take any fat off them, so if they do fly they will not roll down and injure themselves (this of course only applies if they are high quality Birmingham rollers that can perform and they have been locked up for some time). Wheat and Barley and perhaps a little Dari would be my mixture for this purpose.

- When they are first released, ensure they are very very hungry, this probably means that they will not have had food for 48 hours.

- Entice them from the cage to the loft with a conditioned auditory signal of your choice (whistle, voice call, or shaking food can) so that they fly from the cage into the loft or kitbox.

- Offer them a few grains to keep them in close proximity to the loft....that is, allow them to have a walk around the area in front of the loft, or even alight to the loft top. Take care not to give them too much to eat whilst walking around the loft, or you may find them more eager to fly than to go back inside when you decide that the excursion should end.

- Offer them a bath as soon as possible on one of these excursions.

- Ensure that nothing can frighten them whilst this process is taking place.

- An alternative method is to `soap' the wings of valuable old birds and allow them (safe) freedom to walk around the loft area, and on the top of the loft, for a few days before bathing them and removing the soap.

- Another alternative is to use an elastic band around one wing to prevent it from opening and thus keeping the bird from flying. Once again do this only to allow the bird to orientate itself to its new surroundings, and always with its safety in mind, as when it can't fly it is vulnerable to predators.

- Finally a longer term method to prevent flying is to clip the wings of the bird. This is done by cutting the feathers of the primary and secondary flights down the quill on either side (not across the feather - this would lead to bleeding and discomfort for the bird). The bird will not be able to fly to any great extent until it moults out its flights or the flights are pulled to allow regrowth. It will take several weeks for flights to regrow.

23.How long should they fly for?

Flight time will vary according to fitness, age of the roller, and how it has been fed.

Most youngsters will fly for about 30 minutes once trained, and when fit and well fed can fly for an hour or more. Fit and well fed old birds will fly for up to 4 hours, therefore it is important to ensure that they are not overly fit. To ensure this make sure that they are not flown too frequently. A well balanced team of old birds on a liberal measure of food will choose to fly for about an hour and a half. It is most important not to get them too thin, as this will cause high altitude flying and the roller may find it difficult to get down. I believe that many kit losses are due to the birds losing too much body weight and flying too high. If the old bird team begins to fly too high it is important to build more body weight and to fly them less often.

24.My youngsters don't want to fly - what can I do?

Young rollers love to fly, they naturally want to take to the air. When they are reluctant it signifies illness, poor training, or lack of fitness. The most common reason for youngsters not flying is being overweight. The average roller should be between 7 and 10 ounces.

Bring their weight down to the proper flying weight by limiting their food and changing it from wheat to barley or a `depurative' mix available from most commercial sources. Continue to give them access to flight, but don`t expect immediate results, once they are down to a better weight they will usually begin to take to the air. Once the food regime has resulted in a proper weight, if they remain reluctant, then basket them and take them some distance away from the loft and release them. This will get them into the air and begin to get them exercising and building up the required muscles. Take them further and further away until they begin to fly for an appropriate amount of time - perhaps 20 minutes. At this point they should be back on wheat and should begin to want to fly on release. I have on occasions taken a kit one mile away to return them to fitness after being locked up and reluctant to fly, but I would only take a team this far away if it was already properly settled.

One other possible reason for a team of young rollers becoming reluctant, is that they are heading into a heavy moult, they have begun to roll, and they are disturbed by the process and nervous to fly, but generally this will be an individual problem rather than that of a whole team. Of course if the birds are ill, with salmonella, coccidiosis, hair worm infection or endemic canker, a young bird team will be reluctant to fly. Check for symptoms of all these and treat as necessary. Worms are probably the commonest of these ailments, so de-worm as a matter of procedure in any case of reluctance to fly.

25. My youngsters aren't rolling yet - what's the problem?

There may be no problem at all; some youngsters take longer to develop than others. My birds will start to tumble, tail ride, and fast flip at about 5 months old, they will mostly all be rolling reasonably by 9 months, occasionally there may be one or two that take 12

months. The very best of my birds tend to fall into the latter bracket. I have had one bird that miraculously survived without rolling for 4 years. By a series of coincidences: my need for a foster hen - a hen to help settle a young bird team, and also that this bird wouldn't lay but would adopt any eggs and sit them. This allowed her to survive the cull for 4 years, and then she surprised me by beginning to roll in spectacular fashion. This is the exception to the rule though, and generally if the roller isn't rolling by 18 months (you may decide that a bird from a really good pair deserves this amount of time) then it generally isn't going to be worth your while holding on to it. It seems to me that some of the later developing birds turn out to be of very good quality and keep their rolling ability for much longer. It is always good to see a team of old birds that may be 6, 7, or 8 years old still performing to a high standard.
Other reasons for your young rollers not rolling can be their general fitness, they may be too overweight or in poor general condition -although this generally won`t stop them rolling, but might encourage dangerous rolling, bumping, and poor style of roll.

It may be that the birds you have acquired to breed from, or the team of birds you have bought are simply not good Birmingham rollers. If you have acquired them from a bona fide fancier that flies Birmingham rollers in competition, and his or her rollers develop in so many months, then yours should too. Of course it is likely that the birds you have acquired are not of quite the quality of the vendor, but they should be fit for purpose. Contact the vendor and get some advice from the horse's mouth!

To be fair, not every roller sold will be able to produce consistent youngsters that roll. Bob Brown once famously said, 'you have to remember that there isn't a roller in every egg'. However, if you have paid good money for a Birmingham roller you should get a Birmingham roller. The percentage of good rollers from a pair of good breeding rollers should be reasonable. For every 10 youngsters bred from a pair a reasonable expectation would be to have 1 brilliant roller, 5 good rollers, 2 poor rollers, 1 roll down or erratic roller, and 1 that doesn't roll at all. Out of these there may be 1 or 2 that don't kit well, or fly too high, roll too much, roll too deep, roll too short etcetera. Such are rollers. Eventually when you have perfected your breeding, got exactly what you desire in your stock, and know your family of rollers really well, then your statistics will be much better......that might take some time though!

26.How can I get my youngsters to kit?

If your youngsters are well bred they will kit properly without any great effort. Sometimes, especially if they have been left a long time before attempting to train them, they can be a problem. However if you persist in flying them regularly even as often at 3 times a day, they will eventually come together as a kit. The only exception to this is the roller that has been produced by poor quality parents, and with these birds their ability to fly properly in a kit is absent and no amount of flying will help. On occasions youngsters will stop kitting when they begin to come onto the roll. Youngsters that fast flip and lose the kit when they turn are a frequent problem. However, most of these birds, with patience, will eventually kit. The youngster that persists in leaving the kit, rolling and finishing facing away from the team will often not change for the better. It is possible to place these in a well-balanced old bird team if one is available, and sometimes these birds will begin to behave better, however when this fault is demonstrated in one bird it will likely crop up in others if the parents are allowed to continue breeding.

27.My birds fly too long - what can I do?

Birmingham rollers love to fly, so they will fly long times if they are fit. Keeping the team slightly overweight will reduce the length of flight; also ensuring that they are not flown too often will keep them from becoming super fit. Birds that become underweight and super fit will fly too long, so this is to be avoided. One other factor worth considering is how nervous they are. If they have been flagged as youngsters it is possible that they are nervous of the loft top. They may be nervous of cats or other predators which you may not be aware of. My advice here is not to flag youngsters too much so that they feel more comfortable landing on the loft top, ensure that their landing area is as easy as possible for them, and if it is too difficult consider erecting a landing pole - a 20-30 foot metal pole with a wooden cross piece on top for the birds to perch. This will also help the individual roller which might be nervous of landing if it has a tendency to roll when attempting to land.

28. My birds don't fly long enough - what can I do?

Birmingham rollers that don't fly long enough are unfit. This is either due to sickness or being under or overweight. A good kit of old birds can become reluctant to fly if underfed or allowed to become overweight. They should also be checked for common ailments such as coccidiosis, hair worm infestations, canker, or salmonella. Remember also that if birds have been confined for any length of time, they will not be fit enough to fly very long when given liberation. Finally if a team is very active, rolling frequently and especially those teams with deep rollers in them, unless fit they will sometimes go into periods where they are reluctant to fly for very long.

29.Why do my birds keep rolling down?

Once again lack of fitness is usually responsible for this, or poor breeding. Unfortunately some Birmingham rollers are just not stable in the roll and will bump when low because they cannot control the distance they roll to remain safe. Generally these birds are simply not good enough to do anything with, and will need to be disposed of. Birds which are overweight, ill, or undernourished may develop poor control and these are completely correctable. I have a red chequer cock which rolls perfectly when in good condition but deteriorates when its weight or fitness is incorrect, this bird is very helpful to me as when it begins to roll deeper it signifies that I need to check the fitness of the kit, because if he is not right then the rest of the team might need attention.

30. My birds just fly around and don't roll - what can I do?

The treatment will depend on the age of the birds. If we assume that these birds are capable of rolling, then they may be immature and therefore you simply need to be patient. If they have been rolling previously and have stopped rolling or developed a fast or `S' pattern of flight, then they need attention. Youngsters often go through a period of fast flying when the rolling almost completely stops, this is remedied easily by feeding slightly less food and a handful of millet. This will slow the wing beat down, and allow them to begin to roll again, also if an `S' pattern has begun the millet will also help them return to a normal flight pattern. There are other methods of increasing roll in a team and changing the dreaded `S' pattern and this is discussed further in the *50 questions answered for the more experienced roller flier.*

31. What should I feed my youngsters on?

Wean your youngsters on a well balanced mix of peas, wheat, dari and small seeds. Most commercial feed suppliers will have such a mix usually labelled `tumbler mix' or `rear and wean' or similar. For weaning youngsters any mix will do to get them used to eating a variety of grains and seeds but avoid large maize in the mix as this is difficult for youngsters to eat. After a few days of the rollers eating for themselves and before you begin to train them switch to wheat only and continue on this until you have a reason to change, such as conditioning for a competition or dietary change to facilitate weight gain or reduction.

32. What should I feed my old birds on?

Wheat. Good quality wheat is sufficient for flying teams of old birds, until dietary change is needed for some purpose. This is discussed in greater detail in *50 questions answered for the more experienced roller flyer.*

33. What's the best feed for breeders?

I know lots of fanciers that have their preferences for feeds for breeding; the simplest answer to this question is any high protein mix. However, I know fanciers that have successfully reared healthy youngsters on a mix of peas and wheat; tic beans and wheat; peas, tic beans, small maize, and wheat; safflower, peas and wheat; and different types of pigeon or poultry pellets. My only advice here is to ensure that the birds eat all they are fed and are not allowed to pick and throw food around the floor leaving it to be soiled, or worse, attracting rodents. Les Bezance, who switched to pellets a few years ago, recommends feeding breeders twice a day and removing what they leave. This has the benefit of encouraging them to feed full crops to their youngsters twice a day, rather than have them feed small amounts over the day when they squeal for food. This is of course only possible for the fancier that has the luxury of time to do this. Personally I have had to feed using hoppers during the breeding season as I often have to work away from home and cannot feed in such a fashion.

34. Does feeding really make such a difference?

What you feed, and how much you feed your rollers on will have a great effect on them. In my opinion it is almost as important as getting the breeding correct, for even the best rollers can be easily spoiled by improper feeding.

35. What feed or seed will make my rollers roll more?

Barley will make your rollers fly longer, rake more and will reduce weight (this will encourage rolling in kits that are flying in small tight circles). Millet in small doses will reduce the speed of flight and this will encourage a fast flying kit to roll more. Wheat in the right amount will keep your kit in a stable condition once they are rolling. If your team are flying too low, then small amounts of canary seed will help them fly higher and encourage more rolling from a team intimidated by lack of altitude.

36. How much should I feed daily?

This will generally be determined by the weather and how many times they are flown. For years the standard measure for 20 rollers in moderate weather being flown once a day has been a 20oz can of wheat (standard baked bean tin). Personally I believe this to be too little as I prefer my rollers to carry more weight than most other fanciers. In moderate weather I would feed 20 fully developed rollers a tin and a quarter and fly them probably 3 times a week. Youngsters I would feed one and a half tins of wheat if I was flying them once a day, and slightly more if twice a day.

All of this depends on the condition of the birds and the fancier must observe the kit to see how they are responding to the feeding. If the birds are flying too long or too short a time examine them and see if they are underweight or overweight and adjust the feed quantity accordingly.

37. How important is grit?

Grit is very important, it is essential for the birds to digest their food properly. However this needs to be given with caution. Grit freely available to youngsters may result in them gorging themselves on it in the absence of food and this can cause serious injury. Liberal amounts of grit given to old bird hens may encourage them to develop eggs and the consequent problem of rolling out the egg sac in flight. For birds not given liberation or those in the breeding pen both normal and mineral grit should be available at all times.

38. Should I supplement the diet with vitamins?

Fat soluble vitamins A and D are deficient in many of the grains fed to rollers; therefore supplements of these vitamins are really useful to ensure good health and prevention of ill health. Occasional feeds of seeds rich in these vitamins are advisable from time to time. My birds have a small amount of seed most weeks, or in the absence of this for competition purposes I may coat the wheat with a little cod liver oil once a week. Breeding birds must have vitamin supplements during the breeding season to ensure perfect health. Cod liver oil mixed in with feed from time to time would be my preferred method.

39. What should I look for in the feed I buy?

Grain should be dry, hard, and clean. Biting grain in your teeth should be similar to biting a hard biscuit. There should be no signs of infestations with weevil, this shows with tiny holes in the grain, especially noticeable in peas. There should be no chaff or husks in the grain - why pay for inedible material - and definitely no signs of rodent droppings which can cause intestinal obstruction in the bird and consequent death.

40. How many pairs should I start with?

Personally, if I were starting again I would start with the very best 3 hens and 1 cock that I could afford to buy, or could persuade anyone to lend me for 2 seasons. Failing this I would want to breed from 6 pairs of the best I could find and breed as many youngsters as possible out of each pair (10 from each would be ideal) and fly them for a full season. If this was not possible I would buy a kit of youngsters and fly them for a full season and pick out the best 3 hens and 1 cock and use all the rest as feeders.

The simplest answer to this question would be as many high quality pairs as are available, or as many good quality hens and fewer cocks as could be acquired.

41. How do I select pairs to breed with?

Lots of answers to this one, and discussed more fully in *'50 questions answered for the more experienced fancier'*. However there are 2 instant answers to this:

- The best cock to the best hen you have.

- Acquire the best pair you can from a respected and honest Birmingham roller competitor.

Other than that select pairs that suit you, expect the pair to breed something similar to what they themselves represent in the air. Know what sort of roller you prefer and pick only those that live up to that standard. Pick the type of roller you want to reproduce from studying them in the air, don't be misled trying to make a medium deep roller by pairing a deep one to a short one - breeding results are not an average of the two parents, but the qualities and faults from each individual.

42. What is the ideal size for a nest box?

If being confined for the duration of the breeding season the box should be at least 3 foot in length, 15 inches deep and 15 inches high. If the birds are loose in the breeding section of the loft then ideally the boxes would be the same depth and height but need only be 18 inches long.

43. How do I stop cross treading?

Either confine the birds to large breeding boxes for the entire breeding season, or ensure that the breeding section of the loft is big enough to minimise cross treading. The larger the space given to each pair the lower the illegitimacy percentage will be. Levi in his volume *The pigeon* estimates that in an open loft of 4 cubic metres with 6 pairs breeding one can expect to have about 11% illegitimacy, and with additional space this percentage reduces.

44. I have infertile eggs - what can I do?

Infertile eggs are quite common in the first round with pairs of young hens and young cocks. When older cocks are used with young hens the percentage of clear eggs will reduce. Using very old cocks can also be a problem, for although most cocks will continue to `fill' their eggs up to the age of 9 years, after this their reliability will come into question.

Hens can produce fertile eggs well into their 12th year, but once again they become unreliable after the age of 7. To increase the percentage of fertile eggs ensure that the nest boxes have sufficient room for the cock to tread without interruption, and that sufficient vitamins, minerals and grit are available to ensure the birds are in good breeding condition. It is also important that the pairs have been separated for some time before the breeding season commences. Normally breeding cocks and hens should be separated for 2-3 months to ensure both parties are amorous enough to begin breeding with purpose.

When breeding using locked boxes, after the 1st round of youngsters some cocks will lose interest and not tread as effectively or as often as is necessary. This can be remedied by having a cock loose in the breeding pen allowed to fly up to and perch on the outside of the locked nest boxes. This will often provide enough stimuli to keep the locked up cock bird interested in his hen to tread more frequently. Also if possible pairs locked up should be allowed the freedom of the loft from time to time while the other pairs remain in their boxes. Obviously this cannot happen if there is any other cock free in the loft as a hen in egg will squat down for treading by any cock available.

Finally ensure that the birds have enough light to commence effective breeding. It is the amount (time) of daylight that kick starts the breeding cycle. If the days are not long enough then artificial light can be used to begin the cycle - one important point here is that if artificial light is to be used then have timers to bring the lights on during the early morning dark hours and turn off during the normal daylight hours. This will prevent the timer switching off suddenly after dark and leaving hens off the nest when this happens and consequently eggs going cold and the embryos dying in the shell.

45. I have a lot of cocks fighting and breaking eggs - what can I do?

This generally happens when the cocks have not been settled into their boxes before pairing has begun. It is useful to have cocks identify (and yes fight for) their territory before the hens are presented. Cocks should always be locked in with their hens for a few days before releasing them into the open loft, this way both the cocks and hens will identify and protect their nest box a long time before they lay their eggs. So if this problem exists then the best method is to remove the hens and allow the cocks to get their territory firmly established before the hens are then brought back in. On the other hand occasionally it can be a rogue cock which is causing the problem - although this is rare it does happen. Depending on how valuable that particular cock is for breeding it can either be removed from the breeding programme or put in a locked box with its hen for the duration of the season.

46. Can I fly my breeding birds out while they are breeding?

Yes you can, but it is definitely not advisable! Why?

- Breeding pairs will not be in the right flying condition.
- Because of this they are more likely to roll down and misbehave generally.
- They are likely to be your best birds; they will represent a great loss should a bird of prey decide to eat at your table on one of them.
- Hens may be forming an egg and thus may damage their internal organs when rolling, produce an infertile egg, or roll the egg sac out in the process of performing.
- Just flying cocks alone is possible, but dangerous for all the reasons above as once lost it is unlikely the hen will rear the youngsters in the nest alone, and she will not sit eggs without her cock present (or at least in view).

47. What sort of records should I keep?

Individuals tend to vary but I would recommend the following:

- Colour, ring number and sex (I like to have photographs of all my breeding birds used)
- Date 1st egg is laid
- Eye colour - even or odd eyed
- Age when begun to roll
- Graded 1-10 in terms of quality of roll
- Graded 1-10 in terms of frequency of roll
- Graded 1-10 in terms of safety in the roll
- Any special notes:
    - Late in beginning to roll
    - Early in beginning to roll
    - Poor kitter
    - Landing early
    - Mad tumbling before rolling properly
    - Wingy in the roll
    - Late in striking off when a youngster - needed prompting
    - Flies above the kit
    - Erratic when landing (rolls down when attempting to land)
    - Rolls down when released
    - Slow eater
    - Can't fly for rolling

- Exceptional roller - speed
- Exceptional roller - depth

48. How can I breed good rollers?

More of this in *50 questions answered for the more experienced roller flier.*
However here are my best answers:

- Know what a good roller looks like (for you)
- Acquire the best stock you can
- Learn how to feed, train, and fly good Birmingham rollers
- Read this book again
- Test, breed, and experiment every year
- Watch, listen, and ask questions of the most successful fanciers you know
- Take advice from the best fanciers

49. My youngsters are dying in the nest at 10 days old - what is the cause and what can I do?

There are a few reasons for this. They may be any of the following:

- Untreated canker in the parents causing death in squabs at about 10 days
- Hair worm infection - Parents not wormed
- Salmonella or Coccidiosis - Parents not treated
- Cock driving too strongly to the next nest
- Trampled by fighting in the nest box
- Chilled - hen not sitting during cold hours, often caused by
    - feather irritations (red mite, lice, fleas, flies)
    - disturbed from the nest by predator
    - not sitting when the artificial light switched off

50. I have a lot of soft shelled eggs - what is the cause and what can I do?

This is often an indication that the hen is reaching the end of her fertility. However this is also quite common in youngsters in their first season - an indication of their immaturity. The remedy for this is to ensure there is an ample supply of grit, mineral grit, and vitamin supplement. A hen exhibiting this symptom should also be given half a tablet (human adult dose) of calcium and vitamin D daily for 4 days. This is also the treatment for hens that `go off their legs' after laying.

About the author

Graham Dexter has kept Birmingham Rollers since 1964. He has flown in competition since 1972 and has won nearly all of the major trophies. He initiated the National Championship competition in 1981 and devised the constitution and flying rules.

What if I want to know more?

50 Questions for the More Experienced Roller Flier answers the questions that you may have after having grasped the basics. Once again, Graham Dexter draws on almost half a century's experience in keeping rollers to answer common questions in a clear, understandable manner.

What if I want in-depth coverage on the hobby?

Winners with Spinners is the definitive guide to keeping Birmingham Rollers. With detailed discussion of almost every topic to do with Birmingham Roller pigeons, from loft management to psychology, breeding to competing, as well as interviews with some of the greats of the hobby, this book has something for everyone, from beginners to fully-fledged fanciers.

What about the missus?

A Waste of Good Weather by Janice Russell is a light-hearted novel set around the lives of Stan, a Roller Fancier, and his long-suffering wife Evie. From Middlesbrough to Amsterdam, pigeons to diamond-smuggling, this is a story that every fancier's family will want to read and enjoy.

Printed in Great Britain
by Amazon.co.uk, Ltd.,
Marston Gate.